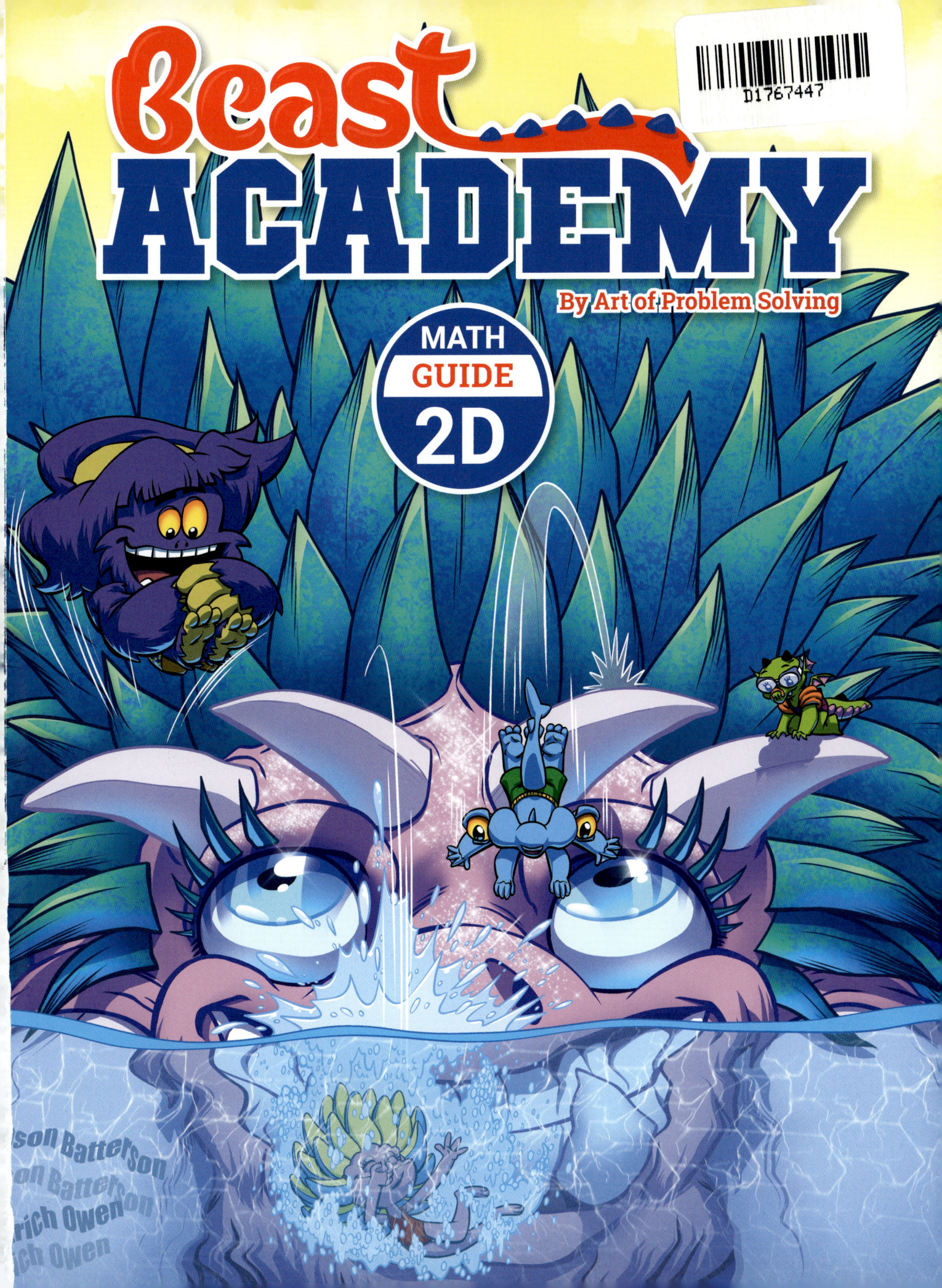

©2019, AoPS Incorporated. All Rights Reserved.

Reproduction of any portion of this book without the written permission of AoPS Incorporated is strictly prohibited, except for fair use or other noncommercial uses as defined in sections 107 and 108 of the U.S. Copyright Act.

Published by:	AoPS Incorporated
		10865 Rancho Bernardo Rd Ste 100
		San Diego, CA 92127-2102
		info@BeastAcademy.com

ISBN: 978-1-934124-36-9

Beast Academy is a registered trademark of AoPS Incorporated.

Written by Jason Batterson
Illustrated by Erich Owen
Additional Illustrations by Paul Cox
Colored by Greta Selman

Visit the Beast Academy website at BeastAcademy.com.
Visit the Art of Problem Solving website at artofproblemsolving.com.
Printed in the United States of America.
First Printing 2019.

Become a Math Beast!
For additional books,
printables, and more, visit
BeastAcademy.com

This is Guide 2D in a four-book series:

Guide 2A
Chapter 1: Place Value
Chapter 2: Comparing
Chapter 3: Addition

Guide 2B
Chapter 4: Subtraction
Chapter 5: Expressions
Chapter 6: Problem Solving

Guide 2C
Chapter 7: Measurement
Chapter 8: Strategies (+&−)
Chapter 9: Odds & Evens

Guide 2D
Chapter 10: Big Numbers
Chapter 11: Algorithms (+&−)
Chapter 12: Problem Solving

Now Available!
Beast Academy Online

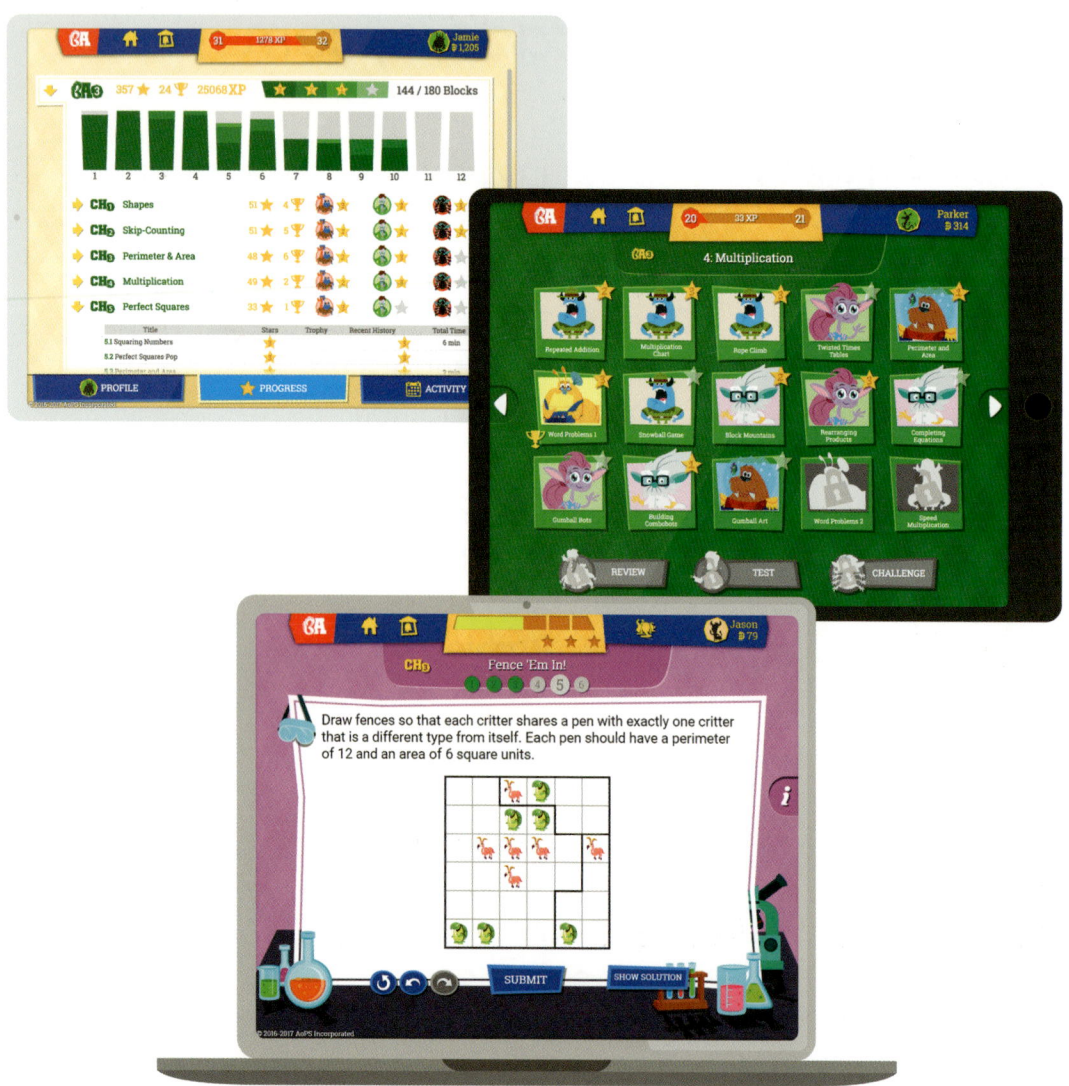

Learn more at BeastAcademy.com

Contents:

Characters 6

How to Use This Book 8

Chapter 10: Big Numbers 12

Thousands and Beyond 14

Grogg's Notes 21

Computing 22

Comparing 28

Estimation 31

Infinity 38

Chapter 11: Algorithms 40

Algorithms 42

Stacking 45

More Than Two 54

Stacking Subtraction 58

Cryptarithms 64

Chapter 12: Problem Solving 74

Counting Paths 76

Organizing 82

Finding a Pattern 89

Math Meet 94

Index 108

Alex
"The Executive"

Irons his socks
Only wears them to bed

Grogg (me!)

I can write with my feet! (not as well as with my hands)

Winnie
"The Firecracker"

Testy at times

Don't be fooled by her cüte handwriting

Lizzie
"The Bookworm"

Read all 52 books in the Dragon Diaries series

Wrote new endings for 3 of them

Contents: Chapter 10

See page 6 in the Practice book for a recommended reading/practice sequence for Chapter 10.

Thousands and Beyond — 14
How do we read the number 100,010,001?

Grogg's Notes — 21
How high would a stack of 1 trillion pennies reach?

Computing — 22
What do you get when you add 43,972+15,000?

Comparing — 28
Which is larger, 1 million or 800 thousand?

Estimation — 31
What are some qualities of a good estimate?

Infinity — 38
Is there a biggest number?

Chapter 10:
Big Numbers

Panel 1:

"5 thousand plus 1 thousand is 6 thousand."

"And adding 1 thousand to 18,000 gives us 19,000."

"We just increase the thousands digit by 1."

5,000 + 1,000 = 6,000
18,000 + 1,000 = 19,000
239,000 + 1,000

Panel 2:

"But when we add 1,000 to 239,000, we can't just increase the thousands digit."

"Adding 1 thousand to 239 thousand gives us 240 thousand."

5,000 + 1,000 = 6,000
18,000 + 1,000 = 19,000
239,000 + 1,000 = 240,000

Panel 3:

"Very good. Let's try adding *more* than 1 thousand."

"How would you add these two numbers?"

43,000 + 15,000

Try it.

23

*THE NUMBER 1,600 CAN ALSO BE READ "SIXTEEN HUNDRED." FOUR-DIGIT NUMBERS THAT END IN EXACTLY TWO ZEROS (1,100 TO 9,900) ARE SOMETIMES READ AS A NUMBER OF HUNDREDS.

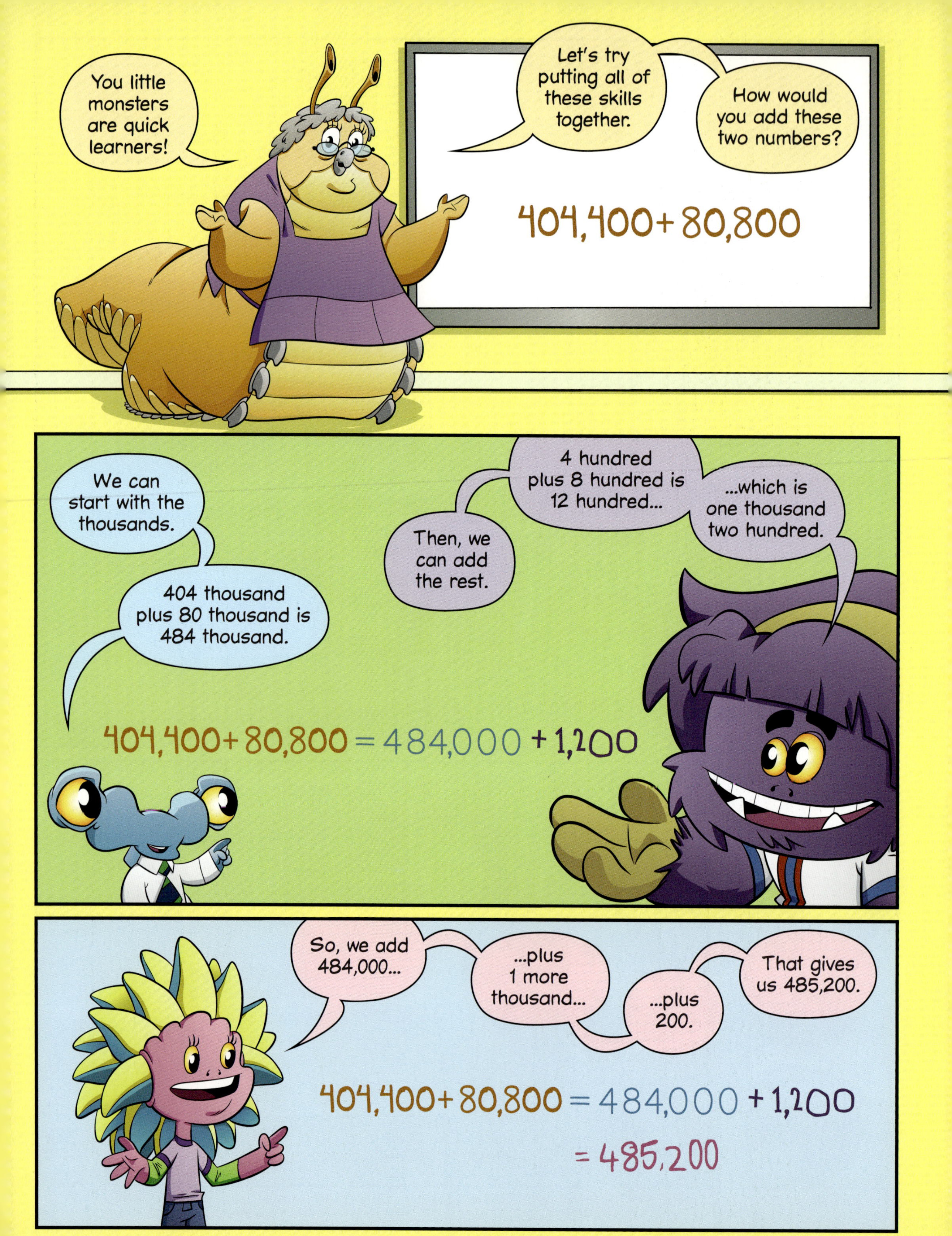

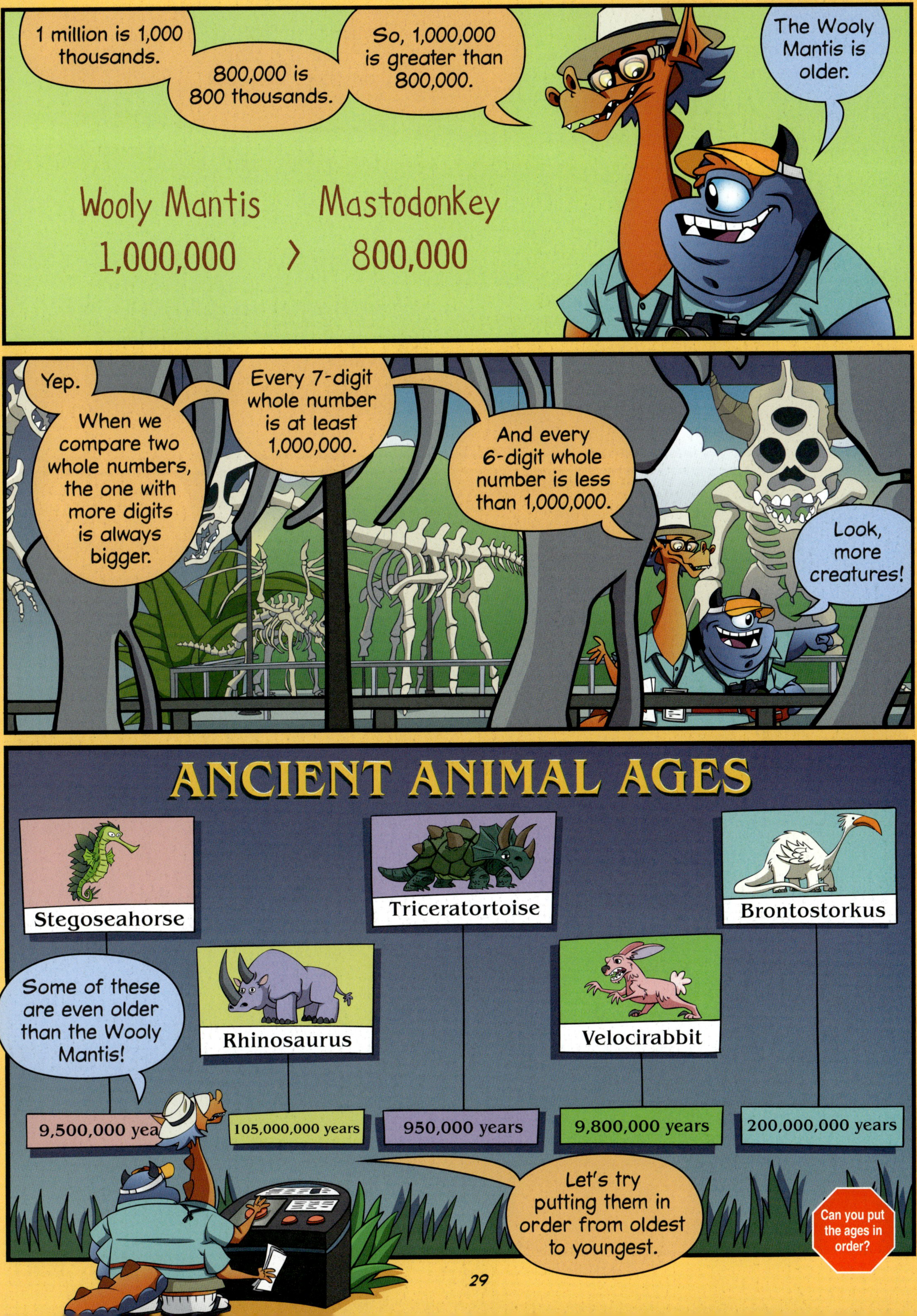

ANCIENT ANIMAL AGES

Stegoseahorse — 9,500,000 years
Rhinosaurus — 105,000,000 years
Triceratortoise — 950,000 years
Velocirabbit — 9,800,000 years
Brontostorkus — 200,000,000 years

Some of these are even older than the Wooly Mantis!

Let's try putting them in order from oldest to youngest.

Can you put the ages in order?

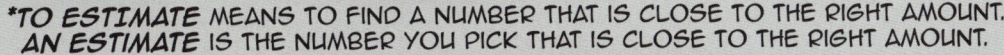

I looked for numbers that are close to the actual numbers, but easier to add.

The sum is about the same as 160,000+90,000, which is 250,000.

$158,921 + 93,162 \approx 160,000 + 90,000$

$160,000 + 90,000 = \boxed{250,000}$

I used 160,000+95,000 to estimate, since 93,162 is closer to 95,000 than to 90,000.

So, the sum is about 160,000+95,000, which is 255,000.

$158,921 + 93,162 \approx 160,000 + 95,000$

$160,000 + 95,000 = \boxed{255,000}$

I added 150,000+100,000.

Since one number got smaller, and the other got bigger by about the same amount...

...158,921+93,162 is about the same as 150,000+100,000, which is 250,000.

$158,921 + 93,162 \approx 150,000 + 100,000$

$150,000 + 100,000 = \boxed{250,000}$

Contents: Chapter 11

See page 38 in the Practice book for a recommended reading/practice sequence for Chapter 11.

Algorithms   42
What's an algorithm?

Stacking   45
Why is it helpful to stack numbers with their place values lined up when adding them?

More Than Two   54
Does stacking work for adding more than two numbers?

Stacking Subtraction   58
How does stacking subtraction work?

Cryptarithms   64
What digits do X, Y, and Z stand for if XX+YY=ZYZ?

R & G Algorithms

"What have you got there?"

"It's a paper airplane book I got at the Museum of Flight!"

"Ooooh... Let's make this one!"

"Maybe we ought to start with an easier plane. That one takes 85 steps!"

"How about this one?"

"Just 4 steps!"

1.
Fold the top-right corner to meet the long edge as shown.

2.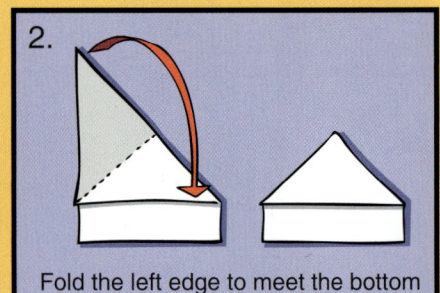
Fold the left edge to meet the bottom edge of the first fold as shown.

3.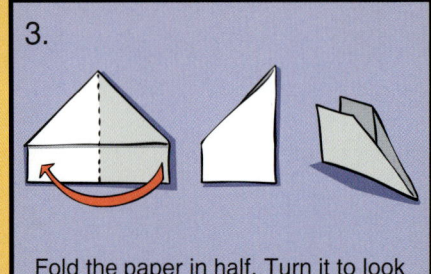
Fold the paper in half. Turn it to look like the diagram on the right.

4.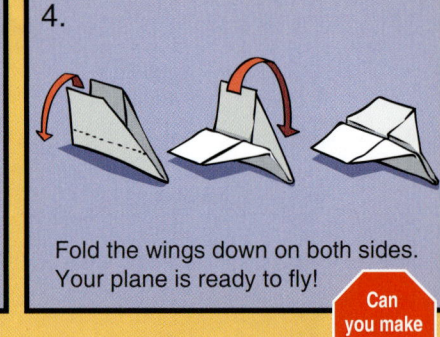
Fold the wings down on both sides. Your plane is ready to fly!

Can you make it?

MATH TEAM: Stacking

"How do we add numbers using place value?"

"We add the tens and the ones separately, then add the results."

$$64 + 78 = 130 + 12 = 142$$

"I found a better way to keep the place values organized."

"How do you do it, Alex?"

"Instead of writing addition in a line..."

"...I like to stack the numbers on top of each other, like this."

$$\begin{array}{r} 64 \\ +78 \\ \hline \end{array}$$

"That's a great idea. Can anyone else see why that's helpful?"

Why is this helpful?

541 + 257

839 + 14,367

777,777 + 99,999

Try all three.

7. Add 345+456+567.

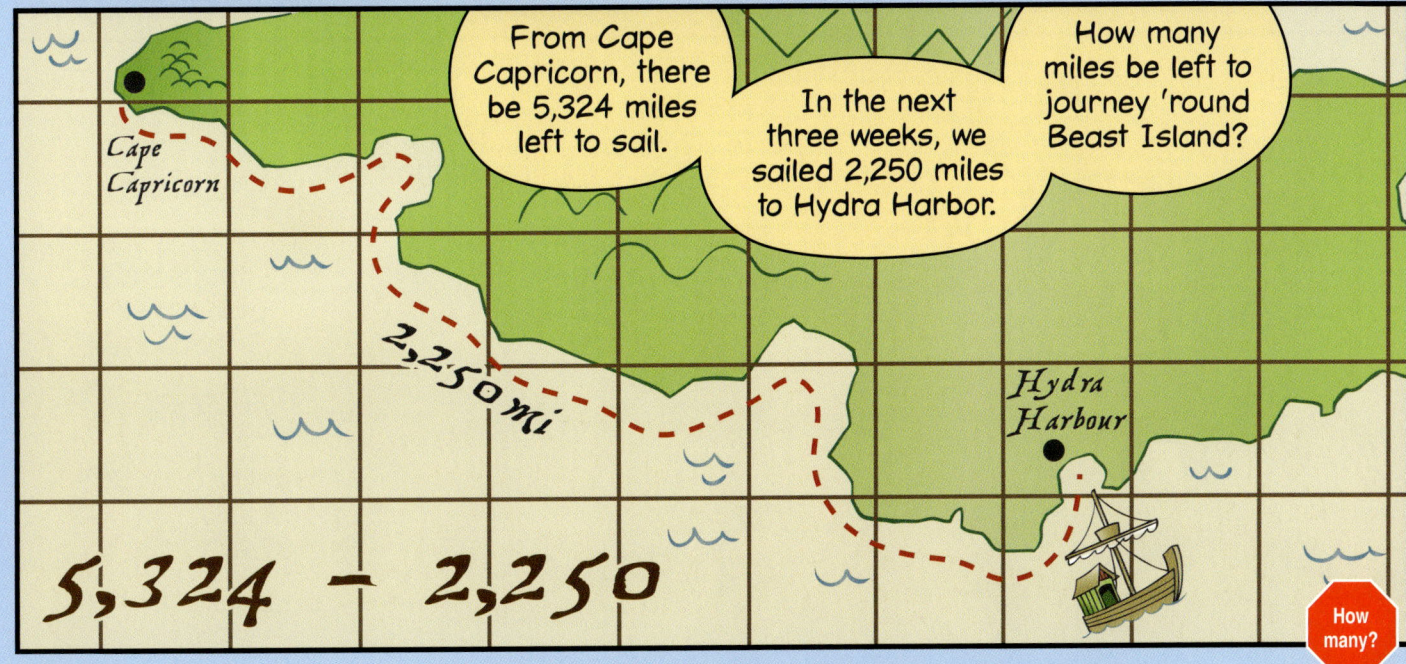

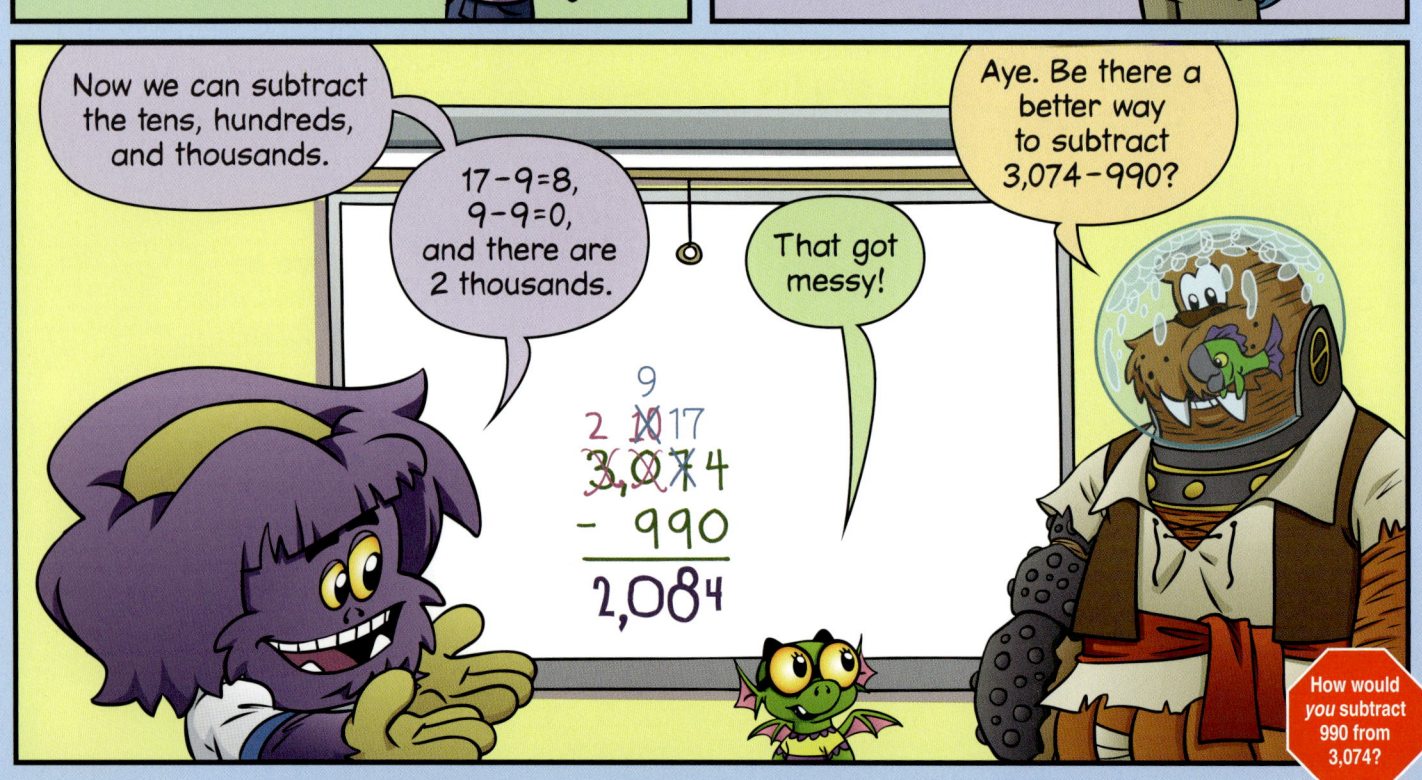

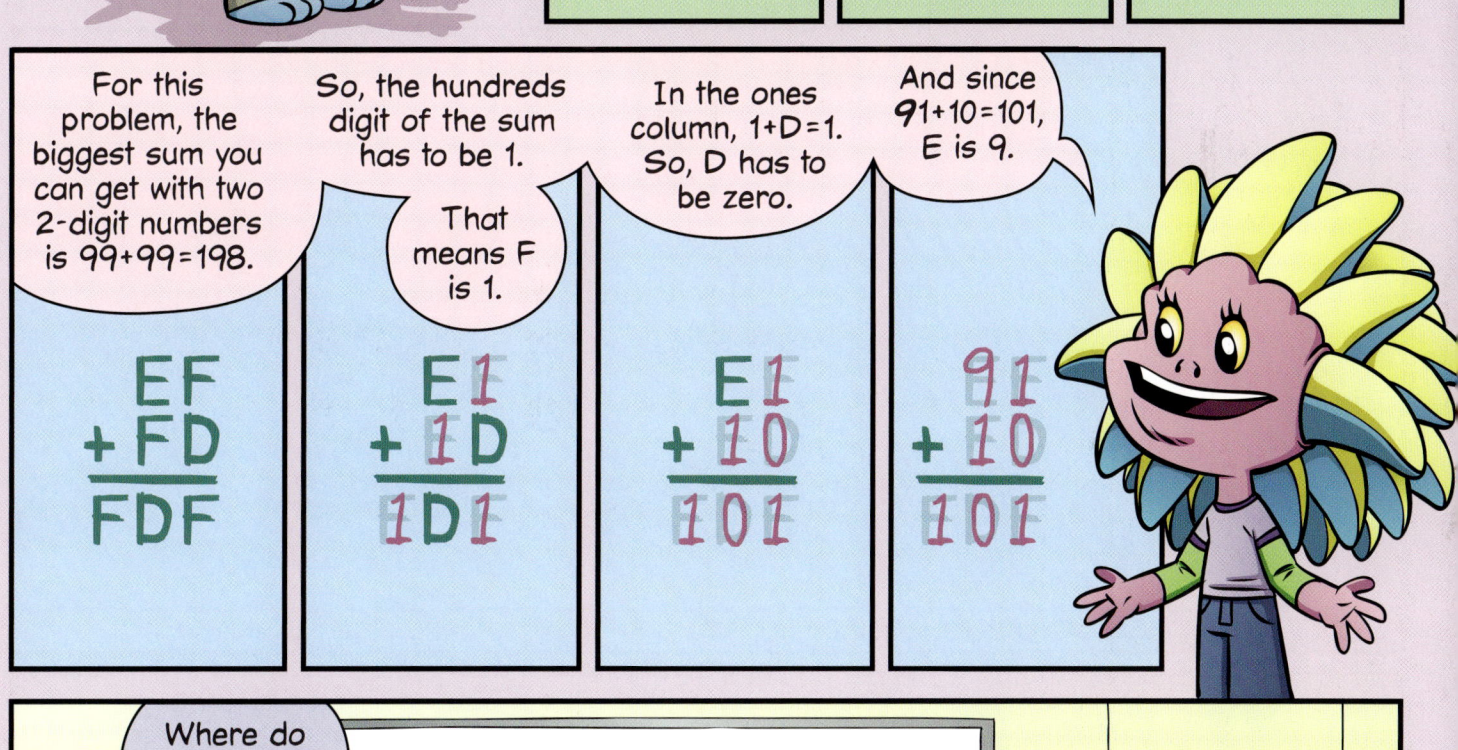

*THERE ARE OTHER NUMBERS THAT END IN 1. WHY CAN'T X+Y EQUAL 21? 31? 41?

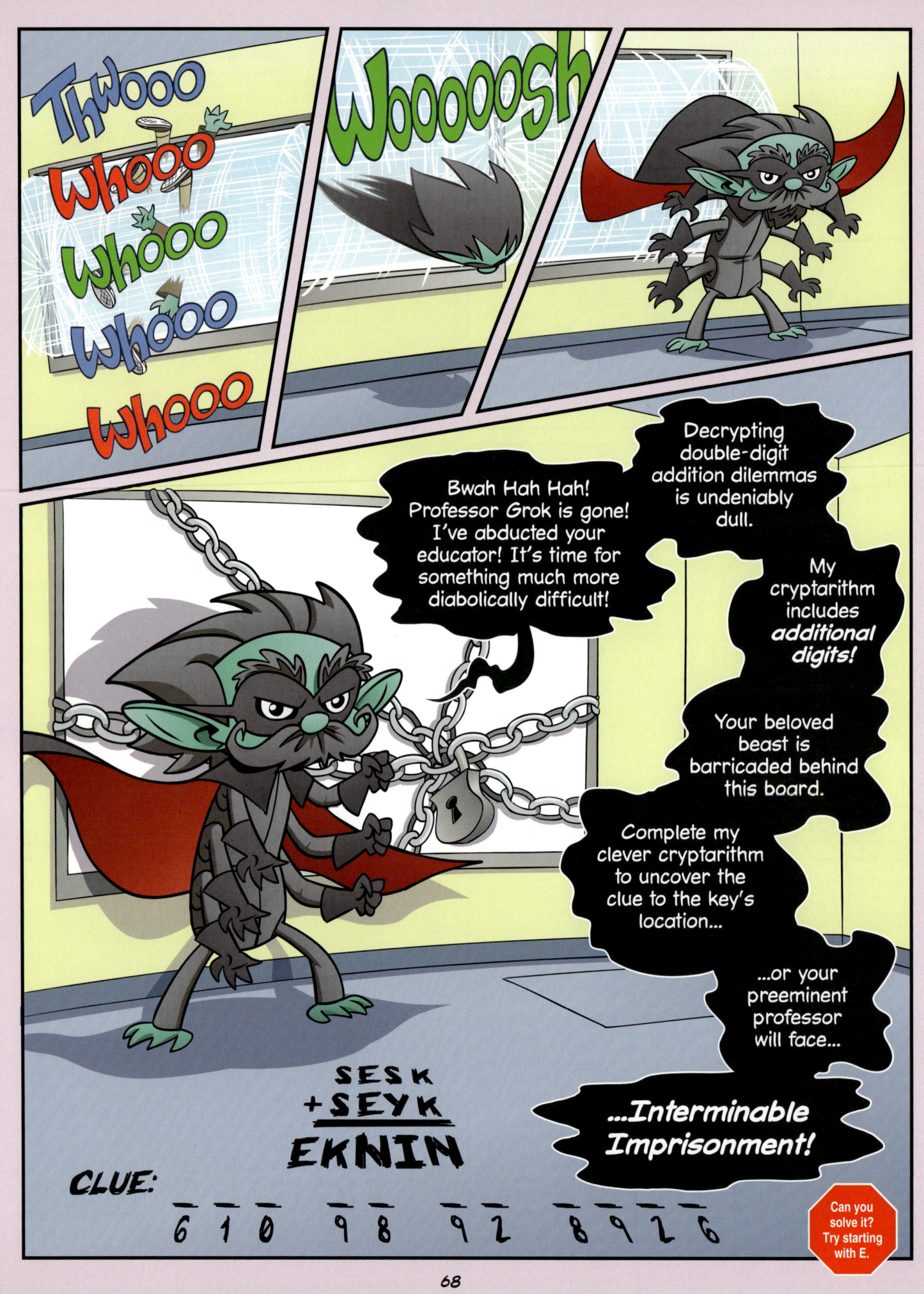

Contents: Chapter 12

See page 70 in the Practice book for a recommended reading/practice sequence for Chapter 12.

Counting Paths 76
How many different routes can Captain Kraken sail between Beast Beach and Cyclops Shore?

Organizing 82
How many different 3-digit numbers can you write using only 3's and 4's?

Finding a Pattern 89
What's the sum of ten 867's?

Math Meet 94
Will the Bots' training help them beat the little monsters, or will Alex, Grogg, Lizzie, and Winnie prevail?

Chapter 12: Problem Solving

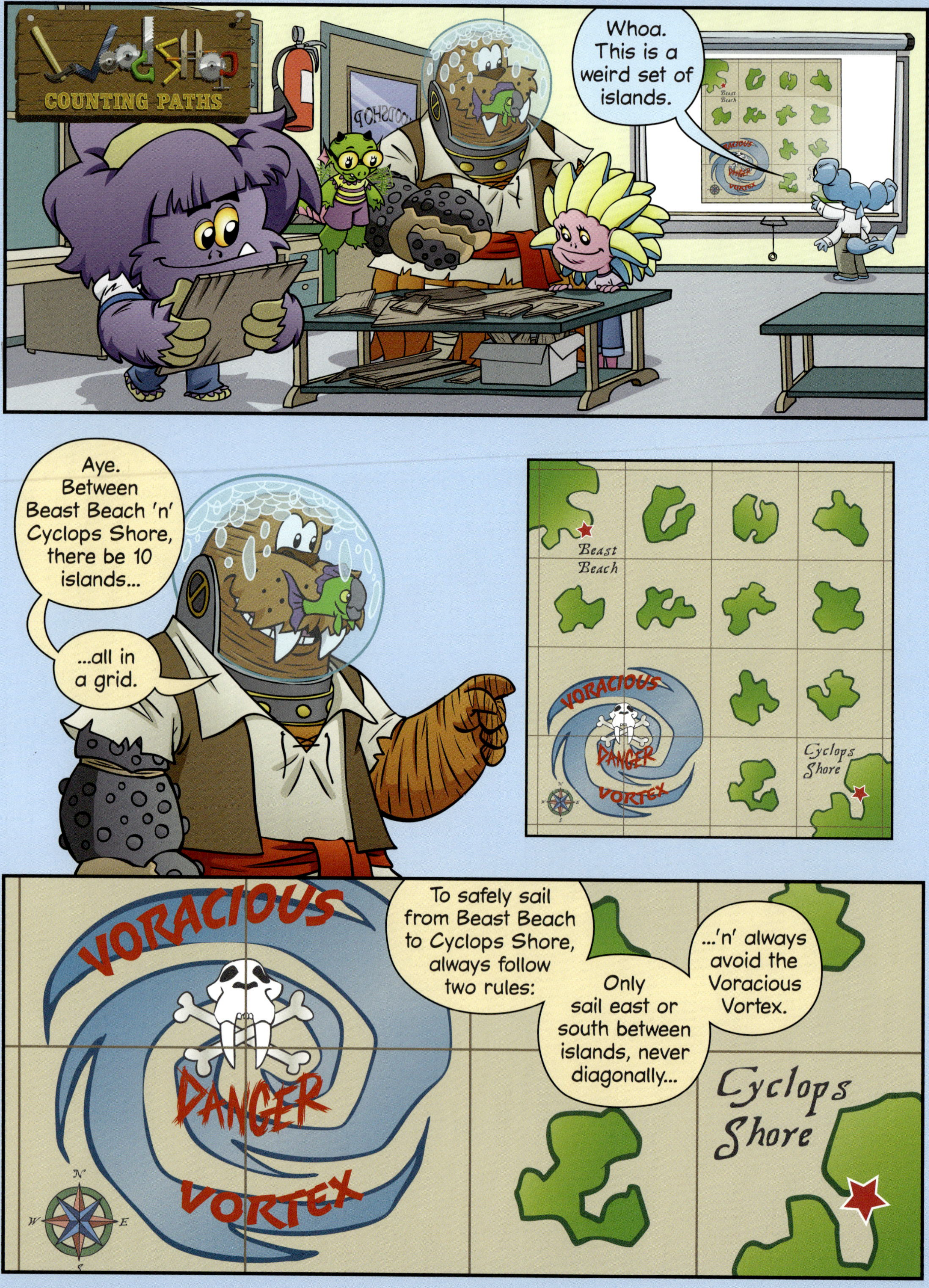

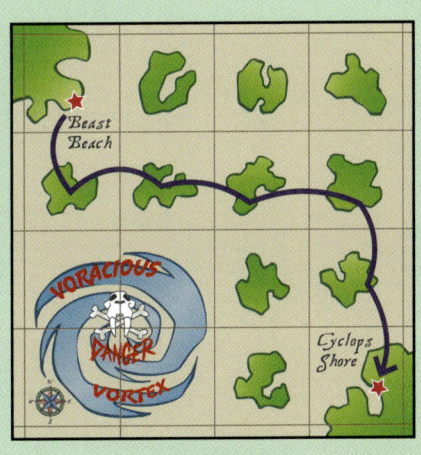

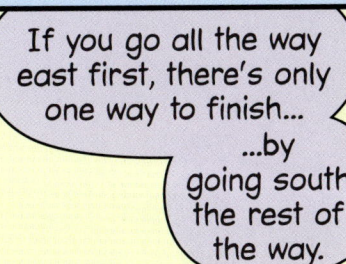

If you go all the way east first, there's only one way to finish... ...by going south the rest of the way.

Great. That's 1 path. Now we can count the paths that start by going east twice before turning south.

Start:	East 3	East 2	East 1	East 0
Paths:	1			

You could go east twice, south *once*, then finish this way...

...or east twice, then south *twice* before turning east...

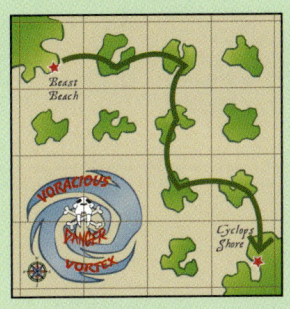

...or you could go east twice, then south *three* times before going east to Cyclops Shore.

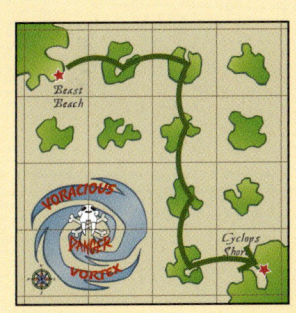

Great. That's 3 more paths. Next, let's count the paths that start by going east once, then south.

Start:	East 3	East 2	East 1	East 0
Paths:	1	3		

How many paths start by going east once?

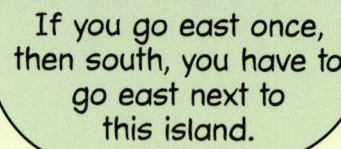

If you go east once, then south, you have to go east next to this island.

There are three ways to finish the trip from there.

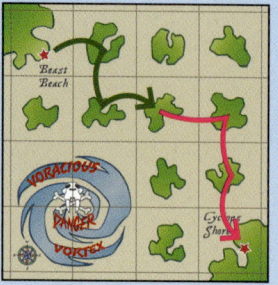

That's 3 more paths.

Now, we need to count the ways you can go if you start by going south instead of east.

Start:	East 3	East 2	East 1	East 0
Paths:	1	3	3	

If you start by going south, you have to go east twice from there.

Or you'll end up in this big, scary, swirly thing!

How many total paths are there?

  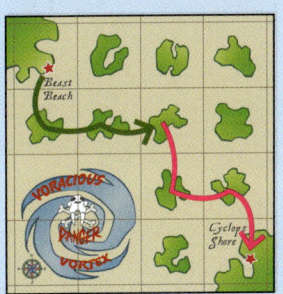 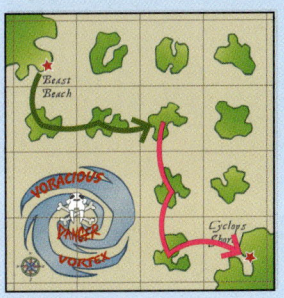

"Just like last time...

"...there are three ways to finish the trip from there."

"That's all of them! That makes a total of 1+3+3+3=10 paths."

"Which one did you take, Captain Kraken?"

Start:	East 3	East 2	East 1	East 0
Paths:	1	3	3	3

"Not the best one, lad...

"...not the best one."

"I started with the only number that has three 3's and no 4's."

"Then, I wrote the numbers that have two 3's and one 4."

"The 4 can go in the ones, tens, or hundreds place."

333 334
     343
     433

"Next, I found three numbers that have one 3 and two 4's."

333 334 443 444
     343 434
     433 344

"The 3 can go in the ones, tens, or hundreds place."

"And finally, there's just one number with three 4's."

"I found 8 numbers, too."

"Excellent! The key to solving many problems like these is being organized."

"How would you organize your work for this problem?"

How many different three-digit numbers have digits that add up to 3?

Try it.

*IF YOU ARE CURIOUS ABOUT THE ACTUAL SUM, FORTY-FIVE 45'S ADD UP TO 2,025.

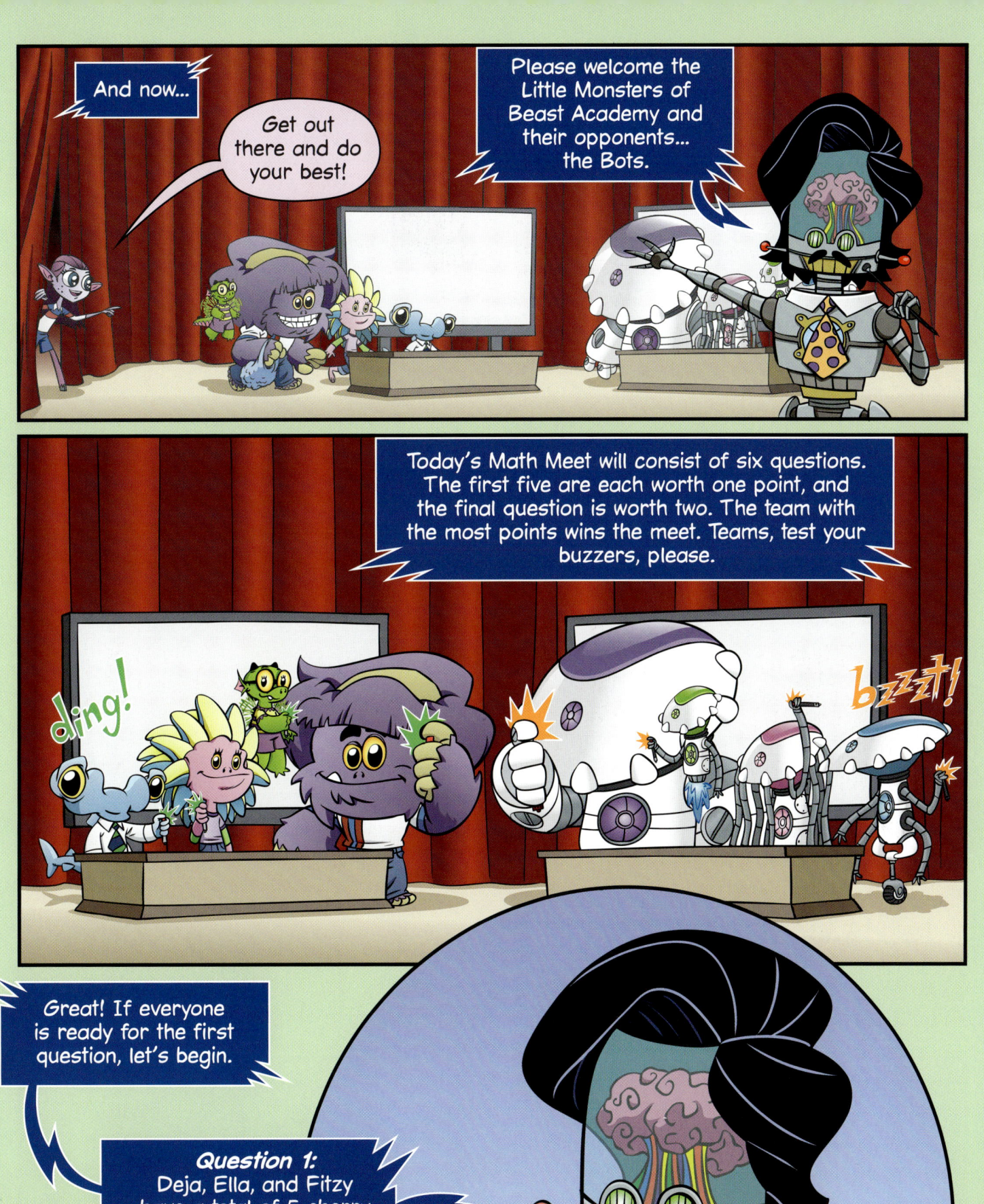

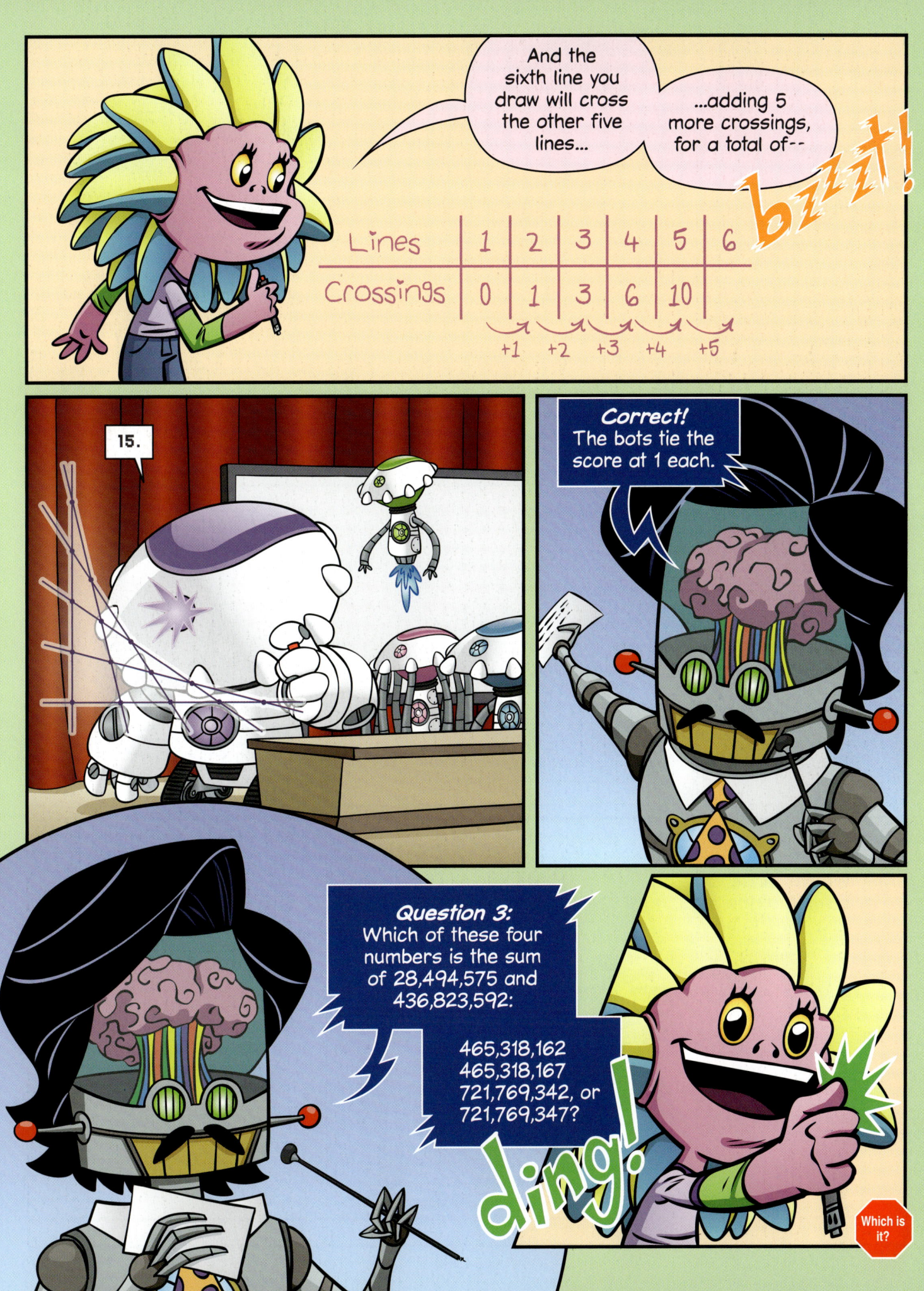

ELIMINATING POSSIBILITIES IS OFTEN A USEFUL TOOL FOR PROBLEM SOLVING.

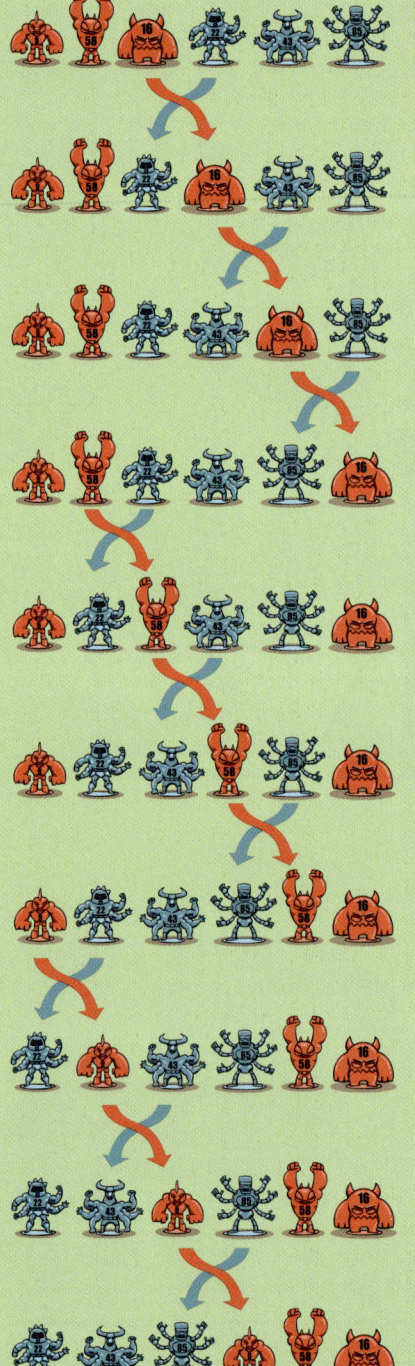

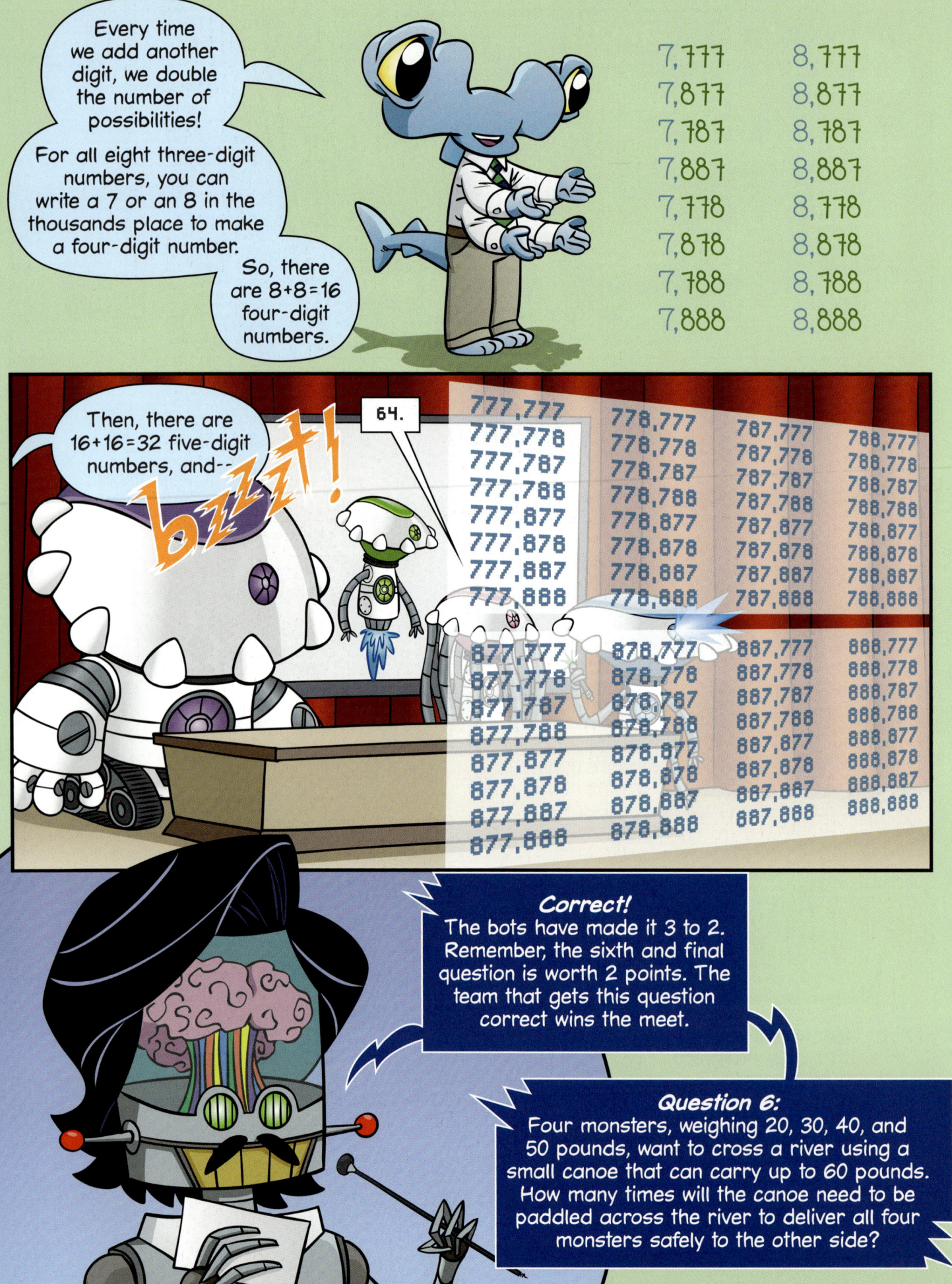

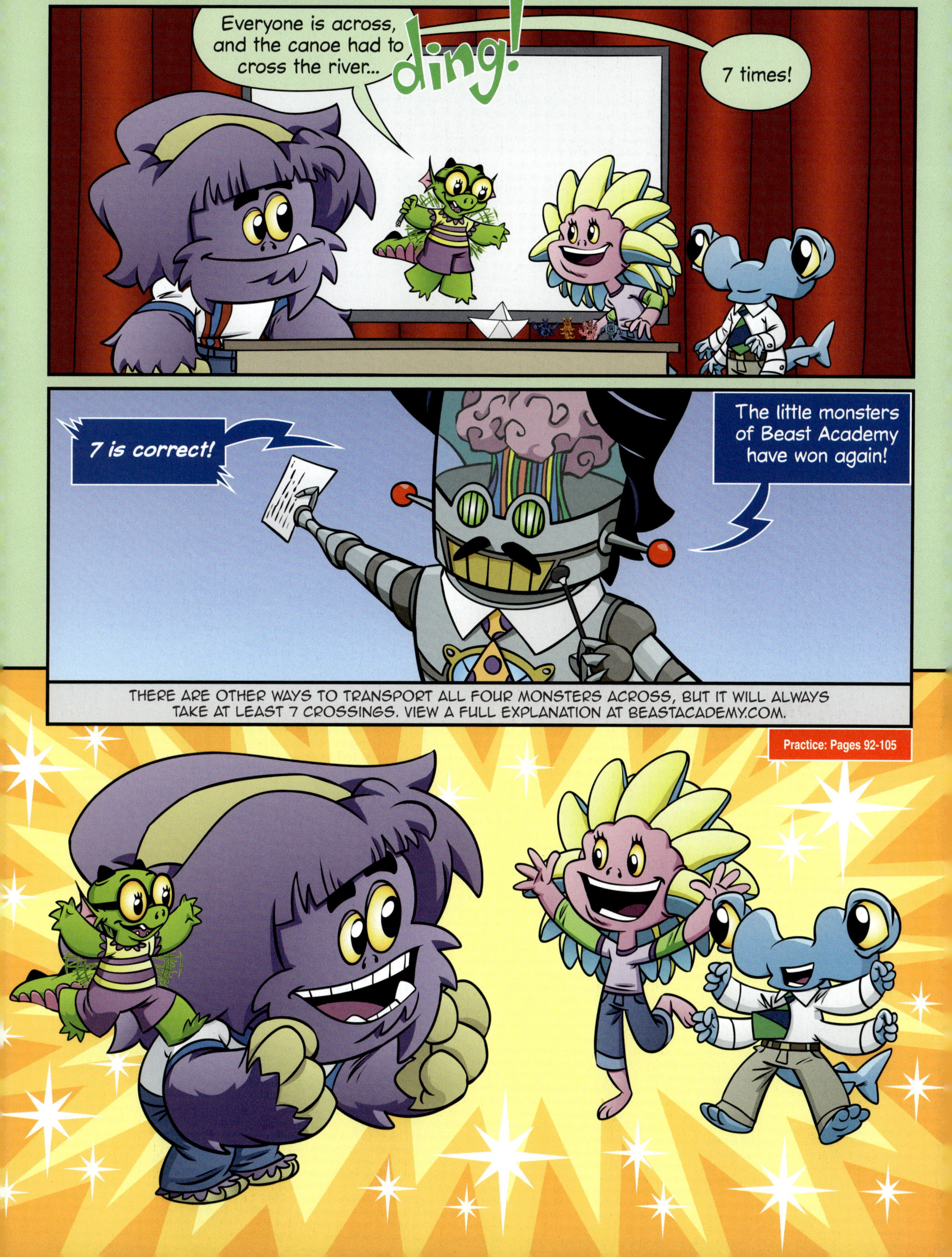

THERE ARE OTHER WAYS TO TRANSPORT ALL FOUR MONSTERS ACROSS, BUT IT WILL ALWAYS TAKE AT LEAST 7 CROSSINGS. VIEW A FULL EXPLANATION AT BEASTACADEMY.COM.

Practice: Pages 92-105

Index

Symbols

≈, 32

A

addition
    adding millions, 25
    adding thousands, 22-24
    estimation, 31-37
    stacking, 45-52, 54-56
    strategies, 50-51
algorithms
    addition, 45-52
    definition, 43
    subtraction, 59-62

B

bazillion, 38
Beast Beach, 76-81
big numbers, 13-39
    commas in, 14, 18
    estimation, 31-37
    reading, 18-19
billion, 19-20
bipeds and biheads, 30
breaking (for subtraction), 60, 62

C

Cape Capricorn, 58-59
Captain Kraken. See Woodshop
chipmonkeys, 101-102
Clara, 27
Clod, Calamitous, 68-69
comparing numbers, 28-30
compubots, 31, 43-44, 52-53
computing, 22-27
    adding millions, 25
    adding thousands, 22-24
Counting by 10's, 90
cryptarithms, 64-73
Cthulhu Cove, 61
Cyclops Shore, 76-81

E

eliminating possibilities, 100
estimation, 31-37

F

finding a pattern, 89-93, 97-99, 103-104
Fiona. See Math Team

G

gajillion, 38
googol, 20
Grok. See Lab

H

Hippopotamoose Lodge, 93
Hydra Harbor, 59, 61

I

index, 108-109
infinity, 38-39

K

Kraken. See Woodshop

L

Lab
    Cryptarithms, 64-73

M

mastodonkey, 28-29
Math Team
    Math Meet, 94-107
    Stacking, 45-53
    Thousands and Beyond, 14-20
Maude the Marauder, 13, 37
Mighty Muscle Monster Minis, 94, 101-102, 105-107
millions, 17-18
Monstro City, 28-30
Ms. Q.
    Computing, 22-27
    Organizing, 82-88

N

Notes
- Grogg's, 21

O

organizing
- lollipops, 95-96
- numbers, 82-85, 103-104
- paths, 78-81
- sums, 85-88

P

penny stacks, 21
place values, 15-19
- millions, 17-18
- stacking, 45-52, 54-56, 58-62
- thousands, 15-17

Playground, 54-57
problem solving
- eliminating possibilities, 100
- finding a pattern, 89-93, 97-99, 103-104
- solving an easier problem, 89-93, 97-99
- using objects, 101-102, 105-106

R

reading big numbers, 18-19, 25
regroup, 22, 25
R&G
- Algorithms, 42-44
- Comparing, 28-30
- Finding a Pattern, 89-93
- Infinity, 38-39

river crossings, 104-107

S

shark week, 88
stacking
- addition, 45-52, 54-56
- subtraction, 58-62

subtraction
- breaking, 60, 62
- stacking, 58-62
- strategies, 62-63

T

tarantulemurs, 101-102
thousands, 15-17
training montage, 53
trillions (and beyond), 19

V

Voracious Vortex, 76-81

W

Woodshop
- Counting Paths, 76-81
- Estimation, 31-37
- Stacking Subtraction, 58-63

wooly mantis, 28-29

For additional books,
printables, and more, visit
BeastAcademy.com